WORK-RELATED LUNG DISEASE SURVEILLANCE REPORT SUPPLEMENT 1992

Division of Respiratory Disease Studies

U.S. DEPARTMENT OF HEALTH AND HUMAN SERVICES
Public Health Service
Centers for Disease Control
National Institute for Occupational Safety and Health
September 1992

DISCLAIMER

Mention of company names or products does not constitute endorsement by the National Institute for Occupational Safety and Health

DHHS (NIOSH) Publication No. 91-113S

Preface

This 1992 supplement to the *Work-Related Lung Disease Surveillance Report* is intended for use with the 1991 *Work-Related Lung Disease Surveillance Report*. The appendix of the 1991 report briefly describes each of the major sources of data used in both the original report and the supplement and directs the reader to additional documentation.

This supplement has two sections: Figures and Tables. Section I contains 16 figures and Section II contains 18 tables. The appendix lists states reporting industry and occupation information to NCHS, and a cross-index by subject is provided.

The 1992 supplement presents updated data for many of the figures and tables presented in the 1991 report. A detailed list of figures and a detailed list of tables provide references to the corresponding figure or table in the 1991 report.

In addition to updated data, this supplement includes data not previously presented. These data include: (1) sex, race, geographic distribution, industry and occupation from the multiple cause of death data for deaths with mention of asbestosis, malignant neoplasms of the pleura, malignant neoplasms of the peritoneum, coal workers' pneumoconiosis, silicosis, byssinosis, or hypersensitivity pneumonitis; (2) number of discharges with silicosis or asbestosis from the National Hospital Discharge Survey; and (3) reports of occupational asthma and silicosis from the Sentinel Event Notification System for Occupational Risks (SENSOR) program.

Surveillance information, including that contained in this report, derives from various sources which differ in completeness of reporting, case definitions, and populations of interest. Nevertheless, surveillance information can be of use in establishing priorities for investigation and intervention, as well as in tracking progress toward the elimination of preventable disease.

Comments and suggestions from users of the report, as well as information about the uses to which it is being put, would be appreciated and will be used to increase the utility of future editions. Comments and suggestions may be sent to:

Work-Related Lung Disease Report
Surveillance Section
Epidemiological Investigations Branch
Division of Respiratory Disease Studies
NIOSH
944 Chestnut Ridge Road
Morgantown, WV 26505

Copies of the 1991 *Work-Related Lung Disease Surveillance Report* and this 1992 supplement may be obtained by calling 1-800-35NIOSH.

Acknowledgements

This report was prepared by the staff of the Surveillance Section, Epidemiological Investigations Branch, Division of Respiratory Disease Studies, National Institute for Occupational Safety and Health. Key contributors included Rochelle B. Althouse, Steven R. Game, Karl Musgrave, Barbara A. Bonnett and Helen S. Montagliani. Gregory R. Wagner, Division Director, and Robert M. Castellan, Branch Chief, provided supervision.

Contents

Preface   iii

Acknowledgments   iv

Section I - Figures

List of Figures   1

Mortality
Asbestosis   2

Malignant Neoplasms of the Pleura   4

Malignant Neoplasms of the Peritoneum   6

Coal Workers' Pneumoconiosis   8

Silicosis   10

Byssinosis   12

Hypersensitivity Pneumonitis   14

Hospitalizations   16
Coal Workers' Pneumoconiosis
Asbestosis
Silicosis

Section II - Tables

List of Tables   18

Mortality   20
Asbestosis
Malignant Neoplasms of the Pleura
Malignant Neoplasms of the Peritoneum
Coal Workers' Pneumoconiosis
Silicosis
Byssinosis
Hypersensitivity Pneumonitis

Coal Workers' X-ray Surveillance Program   22

Occupational Illnesses   22

Dust Diseases of the Lungs   25

Dust Exposure Data   27

Compensation   29

Hospitalizations   30
Coal Workers' Pneumoconiosis
Asbestosis
Silicosis

Addendum
Sentinel Event Notification System   31
for Occupational Risks (SENSOR)
Occupational Asthma
Silicosis

Appendix   36

Cross-Index   37

Figures

Figure 1   Page 2
(See 1991 report Figure 1)
Multiple cause of death listings with any mention of asbestosis, U.S. residents age 15 and over, 1968 to 1988

Figure 2   Page 3
(See 1991 report Figure 1)
Multiple cause of death listings with any mention of asbestosis, U.S. residents age 15 and over, 1988

Figure 3   Page 4
(See 1991 report Figure 2)
Multiple cause of death listings with any mention of malignant neoplasms of the pleura, U.S. residents age 15 and over, 1968 to 1988

Figure 4   Page 5
(See 1991 report Figure 2)
Multiple cause of death listings with any mention of malignant neoplasms of the pleura, U.S. residents age 15 and over, 1988

Figure 5   Page 6
(See 1991 report Figure 3)
Multiple cause of death listings with any mention of malignant neoplasms of the peritoneum, U.S. residents age 15 and over, 1968 to 1988

Figure 6   Page 7
(See 1991 report Figure 3)
Multiple cause of death listings with any mention of malignant neoplasms of the peritoneum, U.S. residents age 15 and over, 1988

Figure 7   Page 8
(See 1991 report Figure 6)
Multiple cause of death listings with any mention of coal workers' pneumoconiosis, U.S. residents age 15 and over, 1968 to 1988

Figure 8   Page 9
(See 1991 report Figure 6)
Multiple cause of death listings with any mention of coal workers' pneumoconiosis, U.S. residents age 15 and over, 1988

Figure 9   Page 10
(See 1991 report Figure 11)
Multiple cause of death listings with any mention of silicosis, U.S. residents age 15 and over, 1968 to 1988

Figure 10   Page 11
(See 1991 report Figure 11)
Multiple cause of death listings with any mention of silicosis, U.S. residents age 15 and over, 1988

Figure 11   Page 12
(See 1991 report Figure 17)
Multiple cause of death listings with any mention of byssinosis, U.S. residents age 15 and over, 1979 to 1988

Figure 12   Page 13
(See 1991 report Figure 17)
Multiple cause of death listings with any mention of byssinosis, U.S. residents age 15 and over, 1988

Figure 13   Page 14
(See 1991 report Figure 18)
Multiple cause of death listings with any mention of hypersensitivity pneumonitis, U.S. residents age 15 and over, 1979 to 1988

Figure 14   Page 15
(See 1991 report Figure 18)
Multiple cause of death listings with any mention of hypersensitivity pneumonitis, U.S. residents age 15 and over, 1988

Figure 15   Page 16
(See 1991 report Figure 9)
Number of discharges with mention of asbestosis, coal workers' pneumoconiosis, or silicosis from short-stay non-Federal hospitals, 1970 to 1989

Figure 16   Page 17
(Not included in 1991 report)
Medicare hospitalizations with mention of asbestosis, coal workers' pneumoconiosis, or silicosis, 1984 to 1989

Figure 1. Multiple Cause of Death Listings With Any Mention of *Asbestosis*, U.S. Residents Age 15 and Over, 1968 to 1988

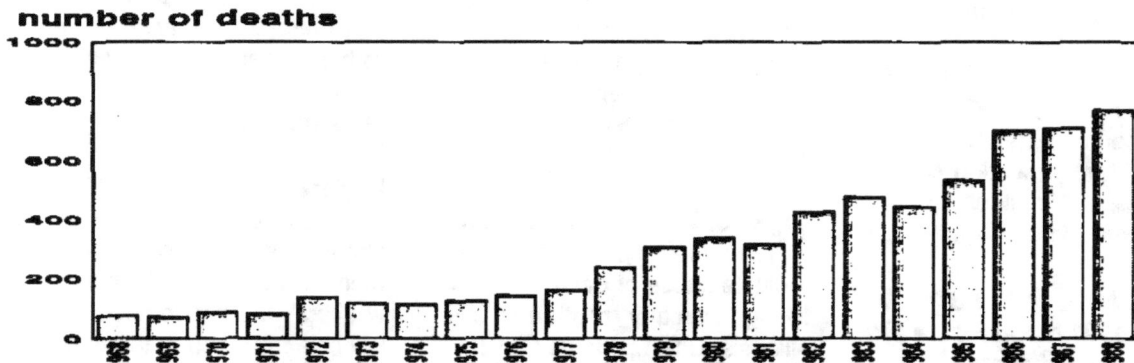

Geographic Distribution

# of Deaths
- 0 to 100
- 101 to 500
- 501 +

Industries Most Frequently Recorded

SIC	Industry	# Deaths
60	Construction	160
360	Ship and Boat Building and Repairing	59
392	Not Specified Manufacturing Industries	17
400	Railroads	15

Occupations Most Frequently Recorded

SOC	Occupation	# Deaths
593	Insulation Workers	55
585	Plumbers, Pipefitters, and Steamfitters	52
19	Managers and Administrators	21
869	Construction Laborers	20

Note: Industry and Occupation Reporting Began in 1985. See Appendix for States Reporting.
Source: NCHS Multiple Cause of Death Tapes

Figure 2. Multiple Cause of Death Listings With Any Mention of *Asbestosis*, U.S. Residents Age 15 and Over, 1988

Distribution by Sex

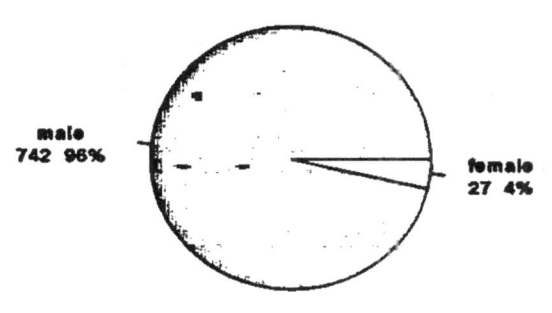

male 742 96%
female 27 4%

Distribution by Race

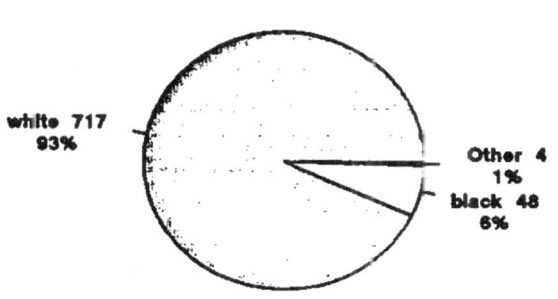

white 717 93%
Other 4 1%
black 48 6%

Geographic Distribution

# of Deaths
- 0 to 25
- 26 to 50
- 51 +

Industries Most Frequently Recorded

SIC	Industry	# Deaths
60	Construction	57
360	Ship and Boat Building and Repairing	21
392	Not Specified Manufacturing Industries	11
192	Industrial and Miscellaneous Chemicals	10

Occupations Most Frequently Recorded

SOC	Occupation	# Deaths
585	Plumbers, Pipefitters, and Steamfitters	19
593	Insulation Workers	15
575	Electricians	11
19	Managers and Administrators	11

Note: 22 States Reported Industry and Occupation in 1988. See Appendix for States Reporting.
Source: NCHS Multiple Cause of Death Tapes

Figure 3. Multiple Cause of Death Listings With Any Mention of *Malignant Neoplasms of the Pleura*, U.S. Residents Age 15 and Over, 1968 to 1988

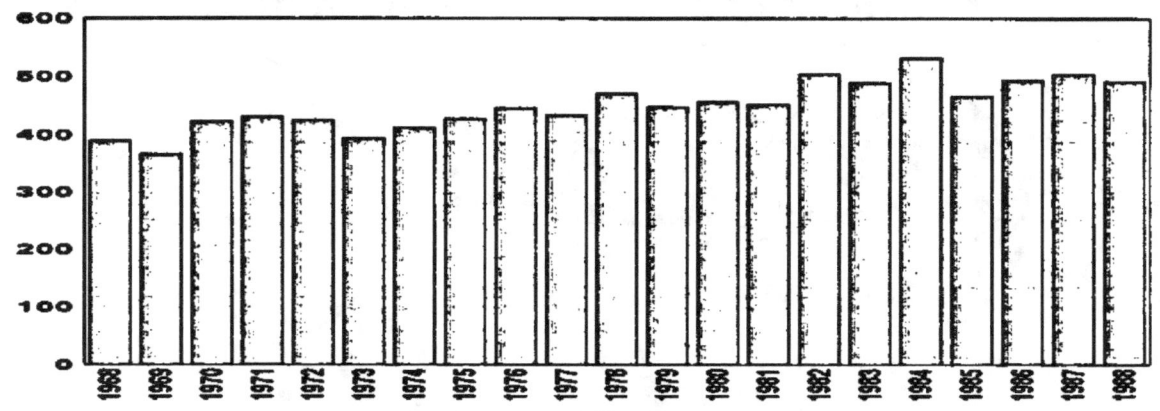

Geographic Distribution

# of Deaths
- 0 to 250
- 251 to 750
- 750+

Industries Most Frequently Recorded

SIC	Industry	# Deaths
961	Homemaker, Student, Unemployed, Volunteer	71
60	Construction	67
400	Transportation, Railroads	14
831	Hospitals	12

Occupations Most Frequently Recorded

SOC	Occupation	# Deaths
914	Homemaker	67
19	Managers and Administrators	29
243	Supervisors and Proprietors, Sales Occup.	16
585	Plumbers, Pipefitters, and Steamfitters	14

Note: Industry and Occupation Reporting Began in 1985. See Appendix for States Reporting.
Source: NCHS Multiple Cause of Death Tapes

Figure 4. Multiple Cause of Death Listings With Any Mention of *Malignant Neoplasms of the Pleura*, U.S. Residents Age 15 and Over, 1988

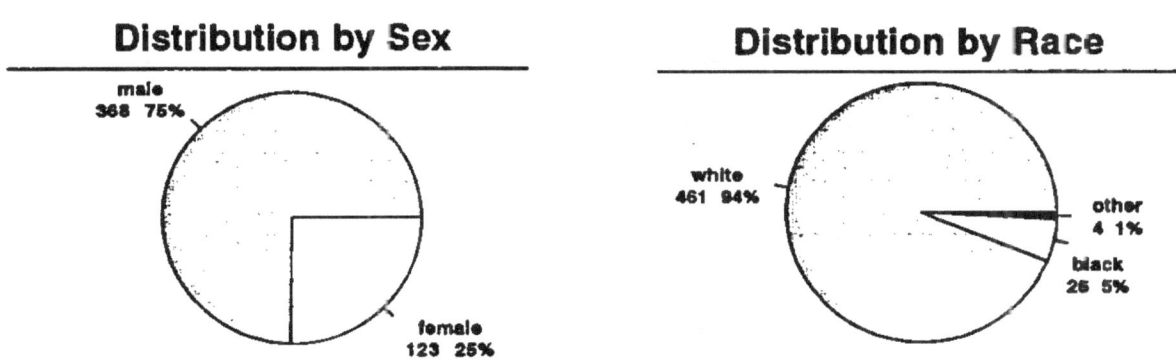

Industries Most Frequently Recorded

SIC	Industry	# Deaths
60	Construction	29
961	Homemaker, Student, Unemployed, Volunteer	22
400	Railroads	6
360	Ship and Boat Building and Repairing	5

Occupations Most Frequently Recorded

SOC	Occupation	# Deaths
914	Homemaker	21
19	Managers and Administrators	7
243	Supervisors and Proprietors, Sales Occup.	7
473	Farmers, Except Horticultural	7

Note: 22 States Reported Industry and Occupation in 1988. See Appendix for States Reporting.
Source: NCHS Multiple Cause of Death Tapes

Figure 5. Multiple Cause of Death Listings With Any Mention of *Malignant Neoplasms of the Peritoneum*, U.S. Residents Age 15 and Over, 1968 to 1988

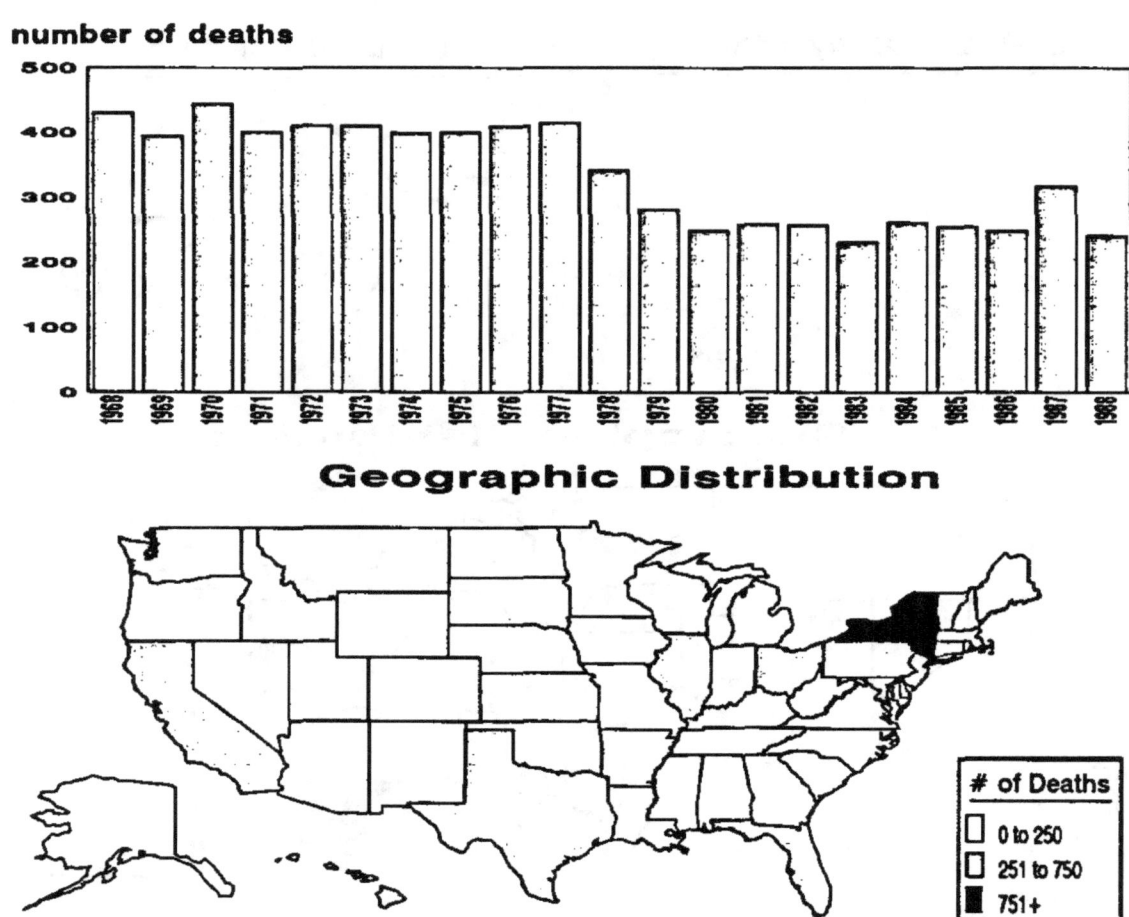

Industries Most Frequently Recorded

SIC	Industry	# Deaths
961	Homemaker, Student, Unemployed, Volunteer	72
60	Construction	19
842	Elementary and Secondary Schools	15
10	Agricultural Production, Crops	9

Occupations Most Frequently Recorded

SOC	Occupation	# Deaths
914	Homemaker	70
473	Farmers, Except Horticultural	11
19	Managers and Administrators	10
156	Teachers, Elementary Schools	9

Note: Industry and Occupation Reporting Began in 1985. See Appendix for States Reporting.
Source: NCHS Multiple Cause of Death Tapes

Figure 6. Multiple Cause of Death Listings With Any Mention of *Malignant Neoplasms of the Peritoneum*, U.S. Residents Age 15 and Over, 1988

Distribution by Sex

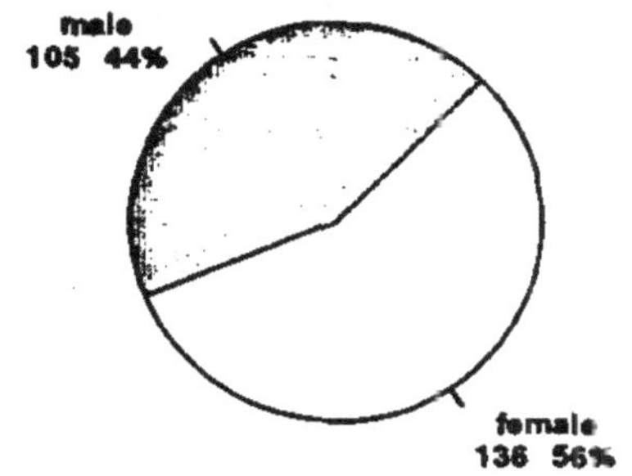

- male: 105 (44%)
- female: 136 (56%)

Distribution by Race

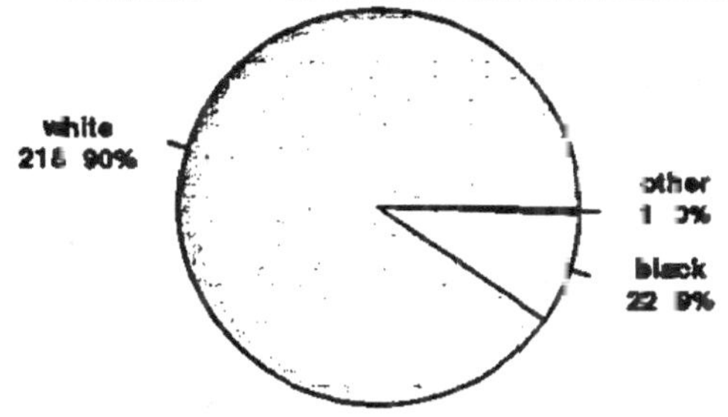

- white: 218 (90%)
- black: 22 (9%)
- other: 1 (0%)

Geographic Distribution

# of Deaths:
- ☐ 0 to 10
- ▨ 11 to 20
- ■ 21+

Industries Most Frequently Recorded

SIC	Industry	# Deaths
961	Homemaker, Student, Unemployed, Volunteer	18
60	Construction	7
842	Elementary and Secondary Schools	4
10	Agricultural Production, Crops	3

Occupations Most Frequently Recorded

SOC	Occupation	# Deaths
914	Homemaker	18
19	Managers and Administrators	4
156	Teachers, Elementary Schools	3
473	Farmers, Except Horticultural	3

Note: 22 States Reported Industry and Occupation in 1988. See Appendix for States Reporting.
Source: NCHS Multiple Cause of Death Data Tapes

Figure 7. Multiple Cause of Death Listings With Any Mention of *Coal Workers' Pneumoconiosis*, U.S. Residents Age 15 and Over, 1968 to 1988

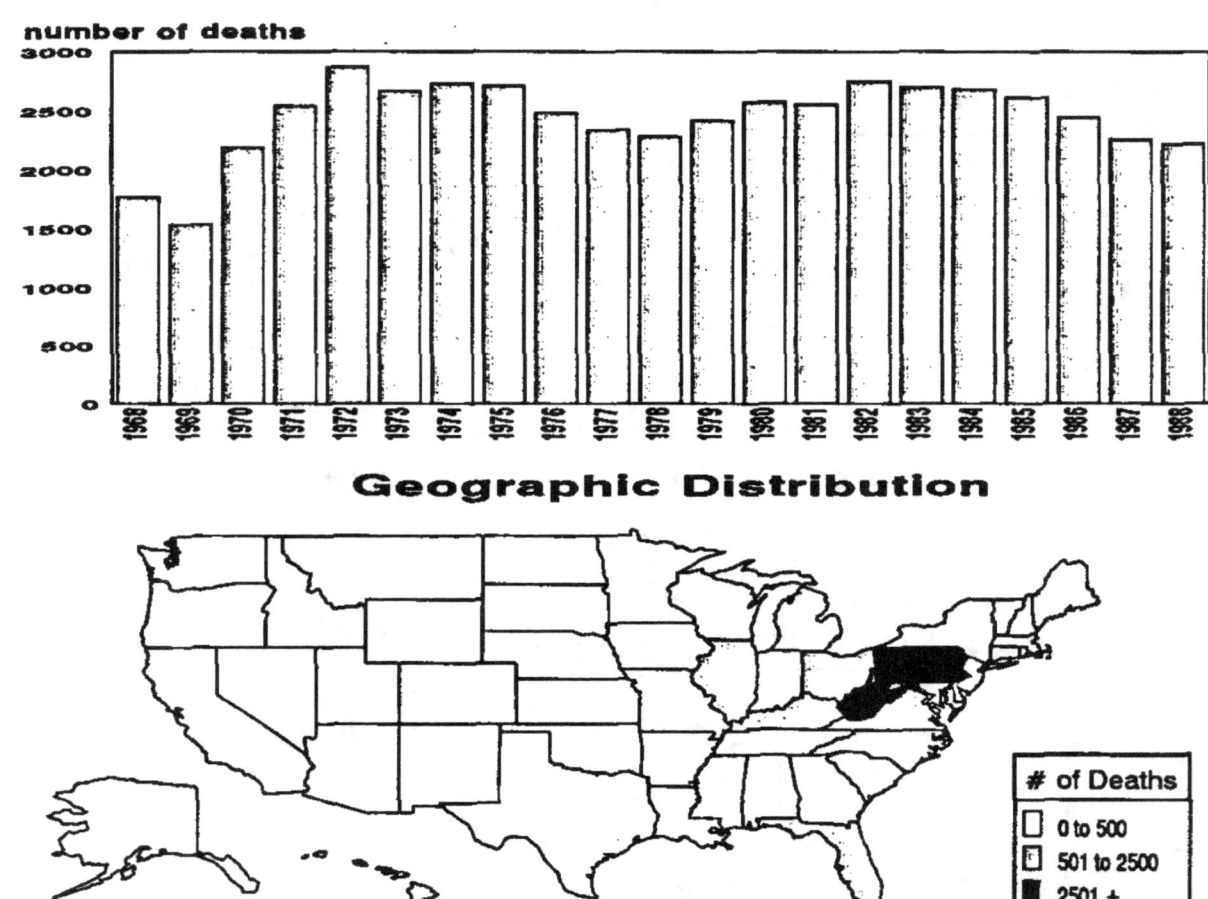

Industries Most Frequently Recorded

SIC	Industry	# Deaths
41	Coal Mining	1406
60	Construction	59
270	Blast Furnaces, Steelworks, Rolling and Finishing Mills	25
392	Not Specific Manufacturing Industries	23

Occupations Most Frequently Recorded

SOC	Occupation	# Deaths
616	Mining Machine Operators	1340
889	Laborers, Except Construction	53
19	Managers and Administrators	22
473	Farmers, Except Horticultural	22

Note: Industry and Occupation Reporting Began in 1985. See Appendix for States Reporting.
Source: NCHS Multiple Cause of Death Tapes

Figure 8. Multiple Cause of Death Listings With Any Mention of *Coal Workers' Pneumoconiosis*, U.S. Residents Age 15 and Over, 1988

Distribution by Sex

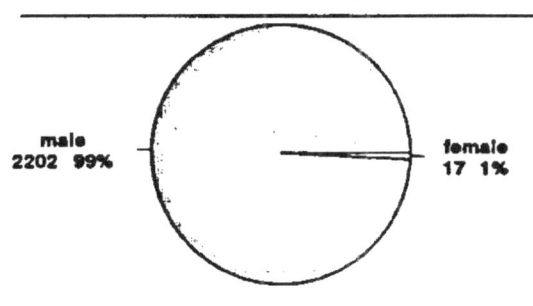

male 2202 99%
female 17 1%

Distribution by Race

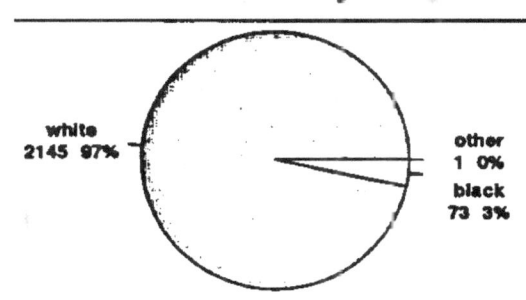

white 2145 97%
other 1 0%
black 73 3%

Geographic Distribution

# of Deaths
- 0 to 50
- 51 to 250
- 251 +

Industries Most Frequently Recorded

SIC	Industry	# Deaths
41	Coal Mining	528
392	Not Specified Manufacturing Industries	13
60	Construction	12
270	Blast Furnaces, Steelworks, Rolling and Finishing Mills	11

Occupations Most Frequently Recorded

SOC	Occupation	# Deaths
616	Mining Machine Operators	501
889	Laborers, Except Construction	19
779	Machine Operators, Not Specified	8
19	Managers and Administrators	7

Note: 22 States Reported Industry and Occupation in 1988. See Appendix for States Reporting.
Source: NCHS Multiple Cause of Death Data Tapes

Figure 9. Multiple Cause of Death Listings With Any Mention of *Silicosis*, U.S. Residents Age 15 and Over, 1968 to 1988

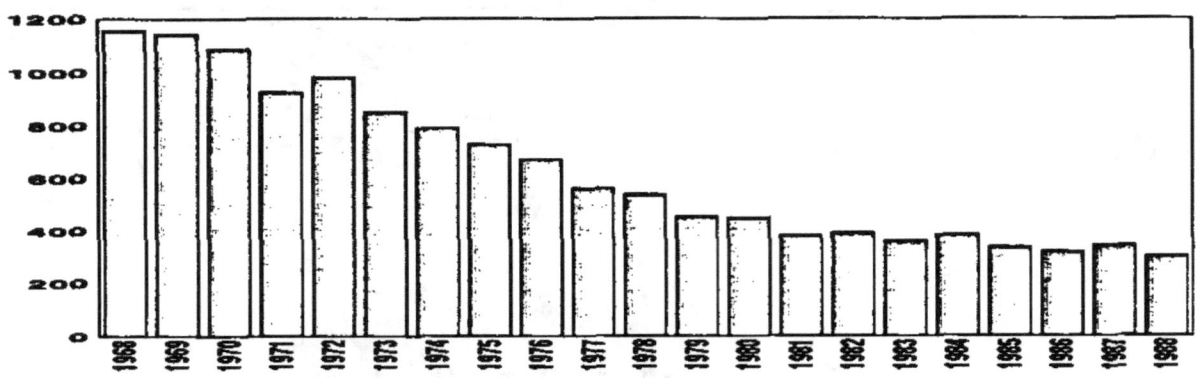

Geographic Distribution

# of Deaths
- 0 to 250
- 251 to 750
- 751 +

Industries Most Frequently Recorded

SIC	Industry	# Deaths
60	Construction	51
270	Blast Furnaces, Steelworks, Rolling and Finishing Mills	31
271	Iron and Steel Foundries	30
262	Miscellaneous Nonmetallic Mineral and Stone Products	29

Occupations Most Frequently Recorded

SOC	Occupation	# Deaths
889	Laborers, Except Construction	63
616	Mining Machine Operators	50
473	Farmers, Except Horticultural	20
19	Managers and Administrators	16

Note: Industry and Occupation Reporting Began in 1985. See Appendix for States Reporting.
Source: NCHS Multiple Cause of Death Tapes

Figure 10. Multiple Cause of Death Listings With Any Mention of *Silicosis*, U.S. Residents Age 15 and Over, 1988

Distribution by Sex

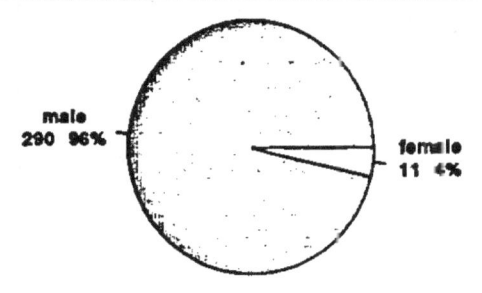

male 290 96%
female 11 4%

Distribution by Race

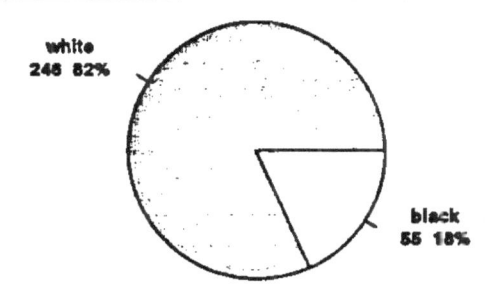

white 246 82%
black 55 18%

Geographic Distribution

# of Deaths
☐ 0 to 15
☐ 16 to 30
■ 31 +

Industries Most Frequently Recorded

SIC	Industry	# Deaths
60	Construction	13
262	Miscellaneous Nonmetallic Mineral and Stone Products	11
40	Metal Mining	9
270	Blast Furnaces, Steelworks, Rolling and Finishing Mills	9

Occupations Most Frequently Recorded

SOC	Occupation	# Deaths
889	Laborers, Except Construction	17
616	Mining Machine Operators	11
473	Farmers, Except Horticultural	9
19	Managers and Administrators	5

Note: 22 States Reported Industry and Occupation in 1988. See Appendix for States Reporting.
Source: NCHS Multiple Cause of Death Tapes.

Figure 11. Multiple Cause of Death Listings With Any Mention of *Byssinosis*, U.S. Residents Age 15 and Over, 1979 to 1988

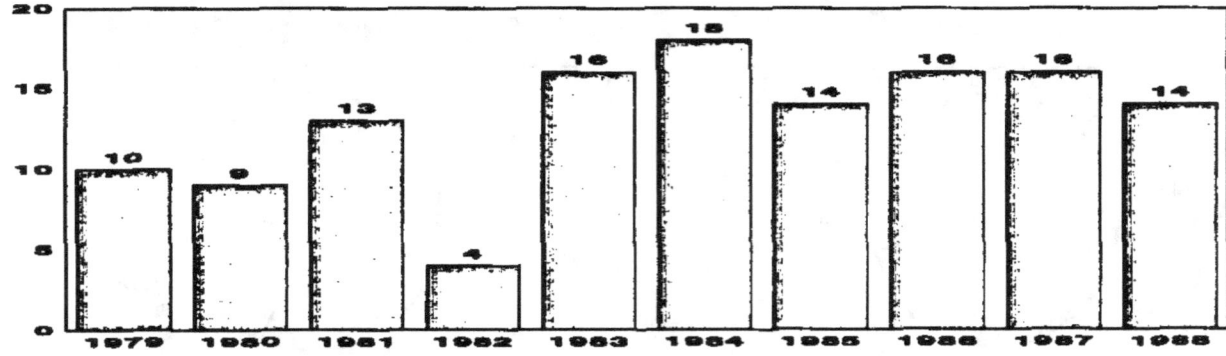

Geographic Distribution

# of Deaths
- 0 to 5
- 6 to 20
- 21 +

Industries Most Frequently Recorded

SIC	Industry	# Deaths
142	Yard, Thread, and Fabric Mills	23
961	Homemaker, Student, Unemployed, Volunteer	2
11	Agricultural Production, Livestock	1
50	Nonmetallic Mining and Quarrying, Except Fuel	1

Occupations Most Frequently Recorded

SOC	Occupation	# Deaths
749	Miscellaneous Textile Machine Operators	10
738	Winding and Twisting Machine Operators	2
779	Machine Operators, Not Specific	2
914	Homemaker	2

Note: Industry and Occupation Reporting Began in 1985. See Appendix for States Reporting.
Source: NCHS Multiple Cause of Death Tapes

Figure 12. Multiple Cause of Death Listings With Any Mention of *Byssinosis*, U.S. Residents Age 15 and Over, 1988

Distribution by Sex

male 12 86%
female 2 14%

Distribution by Race

white 14 100%

Geographic Distribution

# of Deaths
- 0
- 1 to 5
- 6+

Industries Most Frequently Recorded

SIC	Industry	# Deaths
142	Yarn, Thread, and Fabric Mills	8
11	Agricultural Production, Livestock	1
192	Industrial and Miscellaneous Chemicals	1
242	Furniture and Fixtures	1

Occupations Most Frequently Recorded

SOC	Occupation	# Deaths
749	Miscellaneous Textile Machine Operators	3
738	Winding and Twisting Machine Operators	2
779	Machine Operators, Not Specific	2
475	Managers, Farms, Except Horticultural	1

Note: 22 States Reported Industry and Occupation in 1988. See Appendix for States Reporting.
Source: NCHS Multiple Cause of Death Data Tapes

Figure 13. Multiple Cause of Death Listings With Any Mention of *Hypersensitivity Pneumonitis*, U.S. Residents Age 15 and Over, 1979 to 1988

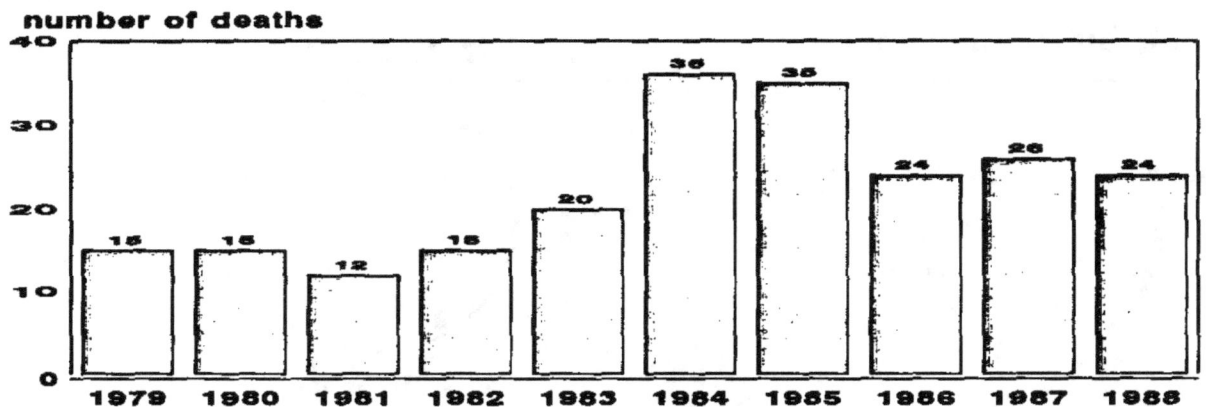

Industries Most Frequently Recorded

SIC	Industry	# Deaths
11	Agricultural Production, Livestock	13
10	Agricultural Production, Crops	11
961	Homemaker, Student, Unemployed, Volunteer	6
150	Miscellaneous Textile Mill Products	1

Occupations Most Frequently Recorded

SOC	Occupation	# Deaths
473	Farmers, Except Horticultural	23
914	Homemaker	6
235	Technicians	1
487	Animal Caretakers, Except Farm	1

Note: Industry and Occupation Reporting Began in 1985. See Appendix for States Reporting.
Source: NCHS Multiple Cause of Death Tapes

Figure 14. Multiple Cause of Death Listings With Any Mention of *Hypersensitivity Pneumonitis*, U.S. Residents Age 15 and Over, 1988

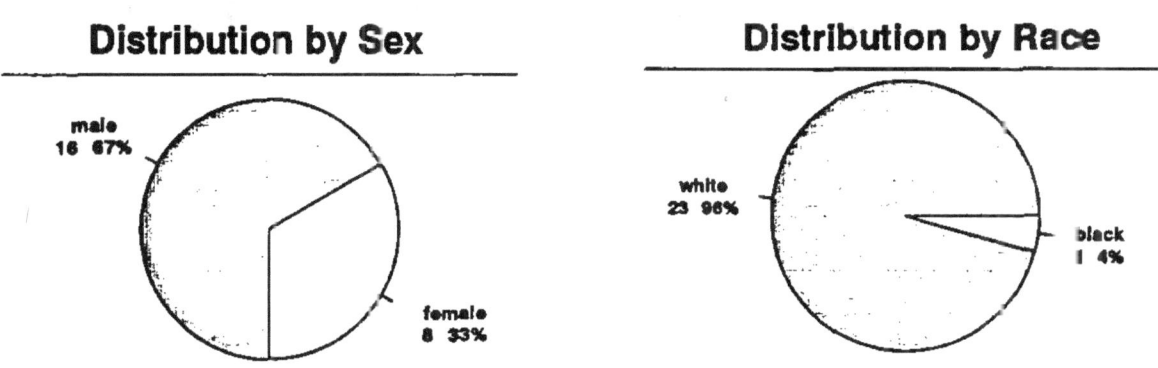

Industries Most Frequently Recorded

SIC	Industry	# Deaths
11	Agricultural Production, Livestock	5
10	Agricultural Production, Crops	2
441	Telephone (Wire and Radio)	1

Occupations Most Frequently Recorded

SOC	Occupation	# Deaths
473	Farmers, Except Horticultural	7
235	Technicians	1

Note: 22 States Reported Industry and Occupation in 1988. See Appendix for States Reporting.
Source: NCHS Multiple Cause of Death Data Tapes

Figure 15. Number of Discharges With Mention of *Asbestosis, CWP, or Silicosis* From Short-stay non-Federal Hospitals, 1970 to 1989

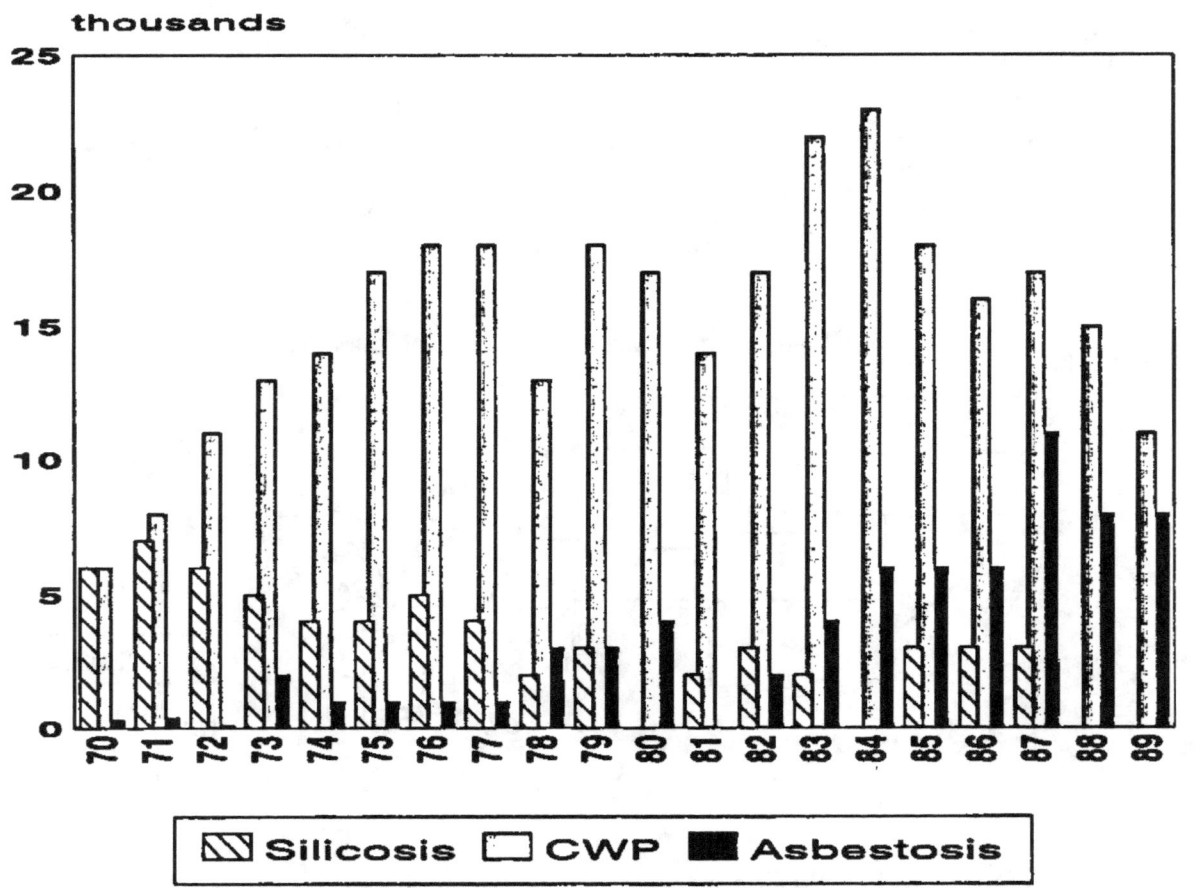

Note: See Table 18 For Data
Source: National Center For Health Statistics

Figure 16. Medicare Hospitalizations With Mention of *Asbestosis, CWP, or Silicosis*, 1984 to 1989

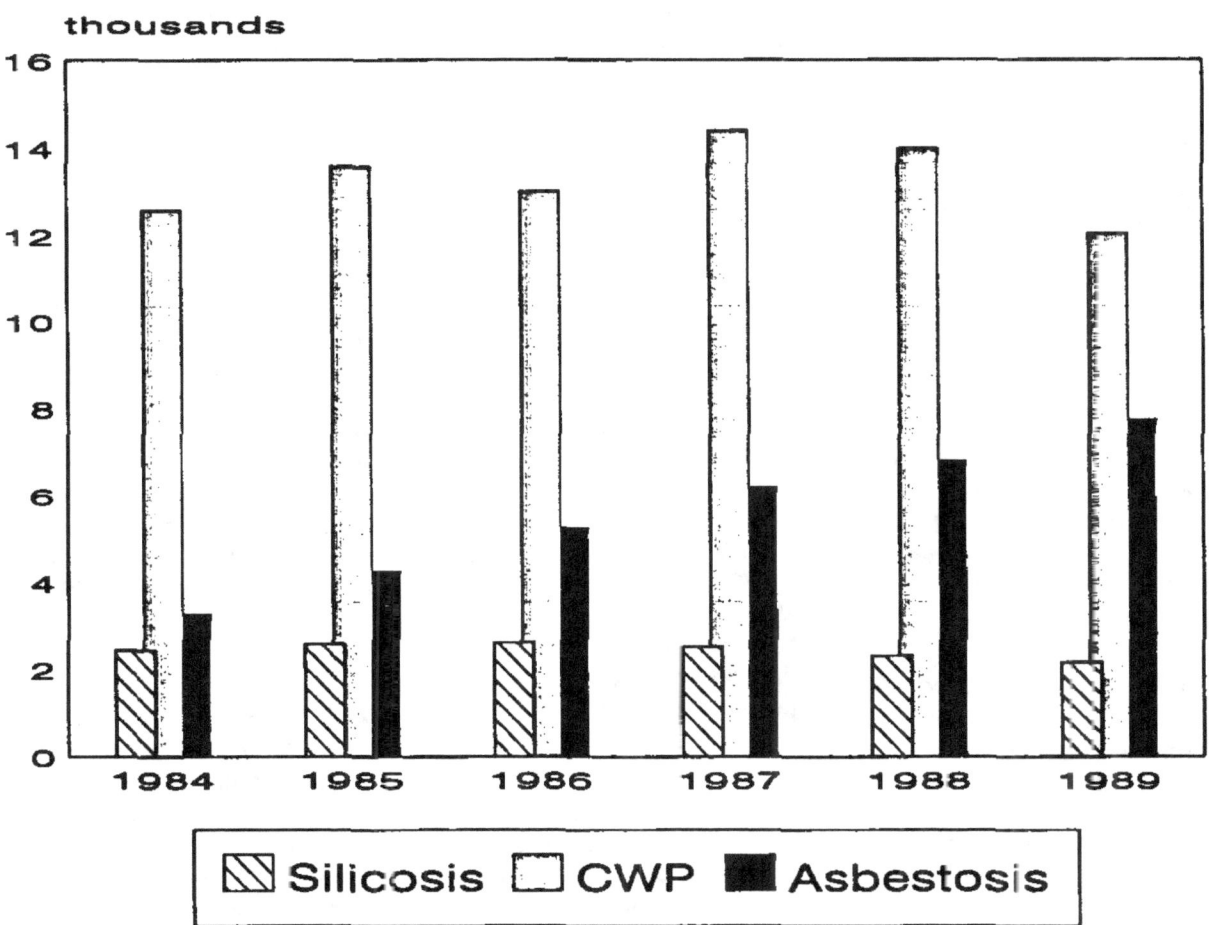

Note: See 1991 Report for Data
Source: Medicare Provider and Analysis and Review, HCFA

Tables

Table 1   Page 20
(See 1991 report Tables 9-11, 20, 28, 34, and 38)
Multiple cause of death listings, United States residents age 15 and over, by state, 1988

Table 2   Page 22
(See 1991 report Table 15)
Number of cases of pneumoconiosis identified in the Coal Workers' X-ray Surveillance Program (CWXSP) by tenure, 1988 to 1990

Table 3   Page 22
(See 1991 report Table 44)
Number of reported occupational illnesses by type of illness for the United States, private sector, 1989 to 1990

Table 4   Page 22
(See 1991 report Table 45)
Percent of reported occupational illnesses by type of illness for the United States, private sector, 1989 to 1990

Table 5   Page 23
(See 1991 report Table 46)
Industries with the largest incidence rates of reported occupational illnesses for the United States, private sector, 1989

Table 6   Page 23
(See 1991 report Table 46)
Industries with the largest incidence rates of reported occupational illnesses for the United States, private sector, 1990

Table 7   Page 24
(See 1991 report Table 47)
Rate per 10,000 full-time workers of reported occupational illnesses by industry division for the United States, private sector, 1989 to 1990

Table 8   Page 24
(See 1991 report Table 48)
Number of reported occupational injuries and illnesses by industry division for the United States, private sector, 1988 to 1990

Table 9   Page 24
(See 1991 report Table 49)
Number of reported occupational illnesses by industry division for the United States, private sector, 1989 to 1990

Table 10   Page 25
(See 1991 report Table 50)
Number of cases of reported occupational dust diseases of the lungs by industry division for the United States, private sector, 1989 to 1990

Table 11   Page 25
(See 1991 report Table 51)
Rate per 10,000 full-time workers of reported occupational dust diseases of the lungs by industry division for the United States, private sector, 1989 to 1990

Table 12   Page 26
(See 1991 report Table 52)
Industries with the highest incidence rates of reported occupational dust diseases of the lungs for the United States, private sector, 1989

Table 13   Page 26
(See 1991 report Table 52)
Industries with the highest incidence rates of reported occupational dust diseases of the lungs for the United States, private sector, 1990

Table 14   Page 27
(See 1991 report Table 54)
Number of dust samples collected by Mine Safety and Health Administration (MSHA) and Occupational Safety and Health Administration (OSHA) inspectors for selected occupational respiratory hazards and the percent of these samples that exceed various levels, 1986 to 1990

Table 15   Page 28
(See 1991 report Table 55)
Number of dust samples collected by Mine Safety and Health Administration (MSHA) and Occupational Safety and Health Administration (OSHA) inspectors for selected occupational respiratory hazards and the percent of these samples that exceed various levels, 1989 to 1990

Table 16   Page 29
(See 1991 report Table 56)
Old Age, Survivors, Disability Insurance (OASDI) Awards for disabled workers with a respiratory diagnosis, by major industry group, 1988 to 1990

Table 17   Page 29
(See 1991 report Table 57)
Number of Black Lung beneficiaries and payments by the Social Security Administration and Department of Labor, 1980 to 1990

Table 18   Page 30
(See 1991 report Table 14)
Number of discharges from short-stay non-Federal hospitals, 1970 to 1988

Table 1. Multiple cause of death listings, United States residents age 15 and over, by state, 1988

State	Asbestosis	Malignant Neoplasm of the Pleura	Malignant Neoplasm of the Peritoneum	Coal Workers' Pneumoconiosis	Silicosis	Byssinosis	Hypersensitivity Pneumonitis
1991 Report Reference Table	Table 9 Page 24	Table 10 Page 25	Table 11 Page 26	Table 20 Page 35	Table 28 Page 43	Table 34 Page 49	Table 38 Page 53
ICD-9 Codes	501	163.0, 163.1, and 163.9	158.8 and 158.9	500	502	504	495
Alabama	10	6	2	33	6	0	0
Alaska	0	0	1	0	0	0	0
Arizona	7	6	2	4	1	0	0
Arkansas	4	6	3	3	0	0	1
California	81	43	12	16	19	1	1
Colorado	6	8	3	21	8	0	0
Connecticut	10	5	4	1	5	0	0
Delaware	8	0	0	2	1	0	0
District of Columbia	0	0	1	1	0	0	0
Florida	59	55	20	35	13	0	0
Georgia	8	5	3	6	6	0	0
Hawaii	3	1	1	0	0	0	0
Idaho	4	1	1	0	4	0	1
Illinois	11	30	18	53	12	0	1
Indiana	5	8	4	25	6	0	0
Iowa	2	6	2	6	2	0	0
Kansas	3	10	2	0	1	0	0
Kentucky	4	10	4	168	11	0	0
Louisiana	12	2	1	0	3	0	1
Maine	12	5	0	0	1	0	1
Maryland	16	4	9	6	2	0	0
Massachusetts	30	10	9	2	5	0	1
Michigan	7	20	8	20	15	0	3
Minnesota	11	15	6	0	8	0	2
Mississippi	15	6	3	2	0	0	0

State						
Missouri	11	5	2	8	2	0	1
Montana	2	3	0	0	6	0	1
Nebraska	2	2	1	0	0	0	1
Nevada	1	0	2	0	1	0	0
New Hampshire	8	1	1	1	2	0	0
New Jersey	83	22	9	24	12	0	1
New Mexico	2	2	0	2	2	0	0
New York	29	45	25	13	12	0	0
North Carolina	12	19	7	6	9	10	1
North Dakota	0	0	1	0	1	0	0
Ohio	12	12	11	102	28	2	1
Oklahoma	5	6	3	4	0	0	0
Oregon	21	18	5	7	2	0	1
Pennsylvania	79	19	14	1153	45	0	2
Rhode Island	11	2	2	1	0	0	0
South Carolina	14	10	1	1	2	0	0
South Dakota	0	0	1	2	1	0	0
Tennessee	12	9	6	45	5	0	0
Texas	46	10	10	7	12	0	0
Utah	1	1	3	16	1	0	0
Vermont	1	2	0	2	2	0	1
Virginia	35	8	6	157	2	0	2
Washington	34	15	5	4	6	0	0
West Virginia	15	6	1	255	5	0	0
Wisconsin	4	12	8	1	16	1	2
Wyoming	1	0	0	4	0	0	0
Total 1988	769	491	241	2219	301	14	24

SOURCE: Tabulations are based on National Center for Health Statistics multiple cause of death data tape for 1988.

Table 2. Number of cases of pneumoconiosis identified in the Coal Workers' X-ray Surveillance Program (CWXSP) by tenure, 1988 to 1990

Tenure (Years in coal mining)	1988			1989			1990		
	X-rays taken	ILO Cat $\geq 1/0$	(%)	X-rays taken	ILO Cat $\geq 1/0$	(%)	X-rays taken	ILO Cat $\geq 1/0$	(%)
0........	371	3	0.8	591	2	0.3	404	3	0.7
1........	78	0	0.0	94	0	0.0	63	0	0.0
2-4.......	221	1	0.4	242	4	1.6	163	0	0.0
5-9........	793	12	1.5	527	6	1.1	261	3	1.1
10-14........	1279	39	3.0	864	26	3.0	441	16	3.6
15-19........	728	39	5.4	447	24	5.4	318	11	3.4
20-24........	229	25	10.9	160	17	10.6	107	13	12.1
25-29........	90	19	21.1	46	10	21.7	31	7	22.6
30+......	135	26	19.2	33	8	24.2	29	5	17.2
Total.......	3924	164	4.2	3004	97	3.2	1817	58	3.2

NOTE: For miners with more than one chest radiograph on file between 1988 and 1990, statistics in this table were calculated based on the most recent x-rays. Tabulations based on data files as of June 30, 1992.

SOURCE: Examination Processing Branch, DRDS, NIOSH.

Table 3. Number of reported occupational illnesses by type of illness for the United States, private sector, 1989 to 1990 (thousands)

Year	Total	Skin diseases or disorders	Dust diseases of the lungs	Respiratory conditions due to toxic agents	Poisoning	Disorders due to physical agents	Associated with repeated trauma	All other occupational illness
1989	283.7	62.1	2.6	18.9	5.8	17.7	146.9	29.7
1990	331.6	60.9	3.0	20.5	6.1	18.2	185.4	37.3

SOURCE: Bureau of Labor Statistics annual reports of occupational injuries and illnesses.

Table 4. Percent of reported occupational illnesses by type of illness for the United States, private sector, 1989 to 1990

Year	Private sector	Skin diseases or disorders	Dust diseases of the lungs	Respiratory conditions due to toxic agents	Poisoning	Disorders due to physical agents	Associated with repeated trauma	All other occupational illness
1989	100.0	22	1	7	2	6	52	10
1990	100.0	18	1	6	2	6	56	11

SOURCE: Bureau of Labor Statistics annual reports of occupational injuries and illnesses.

Table 5. Industries with the largest incidence rates of reported occupational illnesses for the United States, private sector, 1989

Industry	SIC code	Rate per 10,000 full-time workers
Meat Products	201	689.4
Ship and boat building and repairing	373	411.1
Motor vehicles and equipment	371	373.1
Plumbing and heating, except electric	343	346.5
Household appliances	363	275.3
Footwear, except rubber	314	274.7
Motorcycles, bicycles, and parts	375	245.1
Leather tanning and finishing	311	239.9
Rubber and plastics footwear	302	238.6
Flat glass	321	217.9

SOURCE: Bureau of Labor Statistics annual reports of occupational injuries and illnesses.

Table 6. Industries with the largest incidence rates of reported occupational illnesses for the United States, private sector, 1990

Industry	SIC code	Rate per 10,000 full-time workers
Meat Products	201	960.1
Motor vehicles and equipment	371	411.9
Ship and boat building and repairing	373	386.0
Plumbing and heating, except electric	343	354.2
Household appliances	363	346.5
Motorcycles, bicycles, and parts	375	323.6
Public building and related furniture	253	308.6
Hats, caps, and millinery	235	289.7
Forestry	08	272.4
Railroad equipment	374	272.0

SOURCE: Bureau of Labor Statistics annual reports of occupational injuries and illnesses.

Table 7. Rate per 10,000 full-time workers of reported occupational illnesses by industry division for the United States, private sector, 1989 to 1990

Year	Overall	Agriculture	Mining	Construction	Manufacturing	Transportation & Public Utilities	Wholesale & Retail Trade	Finance	Services
1989	37.1	45.5	27.4	17.0	108.3	16.0	9.8	6.1	16.7
1990	43.0	56.4	20.2	18.9	127.7	21.0	10.4	12.4	19.4

SOURCE: Bureau of Labor Statistics annual reports of occupational injuries and illnesses.

Table 8. Number of reported occupational injuries and illnesses by industry division for the United States, private sector, 1988 to 1990 (thousands)

Year	Total	Agriculture	Mining	Construction	Manufacturing	Transportation & Public Utilities	Wholesale & Retail Trade	Finance	Services
1988	6,440.4	101.9	64.4	655.2	2,463.9	464.6	1,533.4	119.5	1,037.6
1989	6,576.3	102.3	60.6	646.5	2,465.5	481.0	1,603.3	118.6	1,098.5
1990	6,753.0	116.3	60.5	638.1	2,429.4	517.9	1,583.8	143.0	1,263.9

NOTE: Because of rounding, components may not add to totals.

SOURCE: Bureau of Labor Statistics annual reports of occupational injuries and illnesses.

Table 9. Number of reported occupational illnesses by industry division for the United States, private sector, 1989 to 1990 (thousands)

Year	Total	Agriculture	Mining	Construction	Manufacturing	Transportation & Public Utilities	Wholesale & Retail Trade	Finance	Services
1989	283.7	4.3	2.0	7.7	204.5	8.4	19.7	3.7	33.6
1990	331.6	5.6	1.5	8.5	235.8	11.4	20.9	7.4	40.5

SOURCE: Bureau of Labor Statistics annual reports of occupational injuries and illnesses.

Table 10. Number of cases of reported occupational dust diseases of the lungs by industry division for the United States, private sector, 1989 to 1990

Year	Total	Agri-culture	Mining	Con-struction	Manu-facturing	Trans-portation & Public Utilities	Wholesale & Retail Trade	Finance	Services
1989	2,600	*	500	200	1,300	100	100	-	200
1990	3,000	100	300	300	1,600	400	100	*	300

NOTE: Because of rounding, components may not add to totals.

SOURCE: Bureau of Labor Statistics annual reports of occupational injuries and illnesses.

* fewer than 50 cases.
- indicates no data reported or data that do not meet publication guidelines.

Table 11. Rate per 10,000 full-time workers of reported occupational dust diseases of the lungs by industry division for the United States, private sector, 1989 to 1990

Year	Total	Agri-culture	Mining	Con-struction	Manu-facturing	Trans-portation & Public Utilities	Wholesale & Retail Trade	Finance	Services
1989	0.3	0.2	7.5	0.5	0.7	0.2	*	-	0 1
1990	0.4	0.6	4.4	0.6	0.9	0.7	-	*	0 1

NOTE: Because of rounding, components may not add to totals.

SOURCE: Bureau of Labor Statistics annual reports of occupational injuries and illnesses.

* incidence rates less than 0.05.
- indicates no data reported or data that do not meet publication guidelines.

Table 12. Industries with the highest incidence rates of reported occupational dust diseases of the lungs for the United States, private sector, 1989

Industry	SIC code	Rates per 10,000 full time workers
Coal mining	12	35.2
Ship and boat building and repairing	373	17.2
Tobacco stemming and redrying	214	9.1
Plastics materials and synthetics	282	4.8
Industrial organic chemicals	286	4.5
Iron and steel foundries	332	4.3
Petroleum refining	291	3.7
Industrial inorganic chemicals	281	2.6
Photographic equipment and supplies	386	2.5
Flat glass	321	2.3
Miscellaneous nonmetallic mineral products	329	2.3

SOURCE: Bureau of Labor Statistics annual reports of occupational injuries and illnesses.

Table 13. Industries with the highest incidence rates of reported occupational dust diseases of the lungs for the United States, private sector, 1990

Industry	SIC code	Rates per 10,000 full time workers
Ship and boat building and repairing	373	33.5
Coal mining	12	20.0
Plastics materials and synthetics	282	5.3
Office furniture	252	3.9
Petroleum refining	291	3.0
Pulp Mills	261	2.7
Fats and Oils	207	2.5
Flat glass	321	2.3
Nonmetallic minerals, except fuels	14	2.3
Highway and street construction	161	2.3

SOURCE: Bureau of Labor Statistics annual reports of occupational injuries and illnesses.

Table 14. Number of dust samples collected by Mine Safety and Health Administration (MSHA) and Occupational Safety and Health Administration (OSHA) inspectors for selected occupational respiratory hazards and the percent of these samples that exceed various levels, 1986 to 1990

Type of Sample	Agency	Total # samples N	Samples < level N (%)	Samples 1-2x level N (%)	Samples > 2x level N (%)	Complaint inspection samples N (%)
Coal Mine Dust						
Surface Mines..................	MSHA	34,589	32,976 (95)	1,208 (4)	405 (1)	-- --
Underground Mines.........	MSHA	70,091	61,377 (88)	7,165 (10)	1,549 (2)	-- --
Quartz Dust						
Coal Mining.....................	MSHA	16,368	11,952 (73)	2,522 (15)	1,894 (12)	-- --
Metal/Non-metal Mining..	MSHA	18,575	13,962 (75)	2,721 (15)	1,892 (10)	-- --
General Industry..............	OSHA					
Level = 10/(%Qrtz+2)		1,713	1,075 (63)	295 (17)	343 (20)	529 (31)
Level = 0.1 mg/m³		975	388 (40)	107 (11)	480 (49)	334 (34)
Asbestos Fiber						
Metal/Non-metal mining..	MSHA	183	181 (98)	1 (1)	1 (1)	-- --
General Industry..............	OSHA					
Level = 2 f/cc		106	105 (99)	1 (1)	0 (0)	46 (43)
Level = 0.2 f/cc		1,001	773 (77)	82 (8)	146 (15)	526 (53)
Cotton Dust	OSHA					
Level = 200 ug/m³		102	57 (56)	27 (26)	18 (18)	42 (41)
Level = 500 ug/m³		5	1 (20)	4 (80)	0 (0)	5 (100)
Level = 750 ug/m³		5	5 (100)	0 (0)	0 (0)	3 (60)
Level = 1 mg/m³		13	9 (70)	1 (8)	3 (22)	1 (8)

NOTE: Levels are defined as follows:

Coal Mine Dust Level = 2 mg/m³ MRE for MSHA coal mine dust sample (level not reduced by quartz content).
Quartz Dust Level = 0.10 mg/m³ MRE for MSHA coal mine quartz dust sample (2 lpm flowrate).
 = 10 mg/m³ divided by (% quartz + 2) for MSHA metal/non-metal mine quartz dust sample (1.7 lpm flowrate).
 = 10 mg/m³ divided by (% quartz + 2) for OSHA quartz dust sample (1.7 lpm flowrate) (1986-March 1, 1989).
 = 0.1 mg/m³ for OSHA quartz dust sample (March 1, 1989-1990).
Asbestos Fiber Level = 2 fiber/cc (8 hours) and 10 fiber/cc (1 hour) for MSHA metal/non-metal mine asbestos sample.
 = 2 fiber/cc for OSHA asbestos sample (1986-June 20, 1986).
 = 0.2 fiber/cc for OSHA asbestos sample (June 20, 1986-1990).
Cotton Dust Level = 200 ug/m³, lint free respirable cotton dust in yarn manufacturing and cotton washing operations; 500 ug/m³, 8 hour TWA, lint-free respirable cotton dust in textile mill waste house operations or lower grade washed cotton in yarn manufacturing; 750 ug/m³, lint-free respirable cotton dust in slashing and weaving processes; and 1 mg/m³, in cotton waste processing operations of waste, recycling (sorting, blending, cleaning, and willowing) and garnetting.

Source: Tabulations by Environmental Investigations Branch, DRDS, NIOSH from data tapes provided by OSHA and MSHA.

-- indicates data not available.

Table 15. Number of dust samples collected by Mine Safety and Health Administration (MSHA) and Occupational Safety and Health Administration (OSHA) inspectors for selected occupational respiratory hazards and the percent of these samples that exceed various levels, 1989 to 1990

Type of Sample	Agency	Year	Total # Samples N	Samples < level N	(%)	Samples 1-2x level N	(%)	Samples > 2x level N	(%)	Complaint Inspection samples N	(%)
Coal Mine Dust											
Surface Mines..	MSHA	1989	6,673	6,401	(96)	192	(3)	80	(1)	–	–
		1990	6,704	6,429	(96)	214	(3)	62	(1)	–	–
Underground Mines.............	MSHA	1989	13,306	11,721	(88)	1,329	(10)	256	(2)	–	–
		1990	12,007	10,759	(90)	1,020	(8)	228	(2)	–	–
Quartz Dust											
Coal Mining...........	MSHA	1989	2,945	2,113	(72)	487	(16)	345	(12)	–	–
		1990	2,698	2,003	(72)	435	(15)	360	(13)	–	–
Metal/Non-Metal Mining	MSHA	1989	4,082	2,901	(71)	677	(17)	504	(12)	–	–
		1990	4,695	3,563	(76)	674	(14)	458	(10)	–	–
General Industry..............	OSHA	1989	588	368	(63)	108	(18)	112	(19)	183	(31)
		1990	484	259	(54)	66	(13)	159	(33)	189	(39)
Asbestos Dust											
Metal/Non-Metal Mining.................	MSHA	1989	25	25	(100)	0	(0)	0	(0)	–	–
		1990	44	42	(95)	1	(2)	1	(2)	–	–
General Industry..............	OSHA	1989	196	151	(77)	16	(8)	29	(15)	98	(50)
		1990	114	90	(79)	7	(6)	17	(15)	59	(52)
Cotton Dust											
Level = 200 ug/m^3	OSHA	1989	6	3	(50)	3	(50)	0	(0)	3	(50)
		1990	17	13	(77)	4	(23)	0	(0)	16	(94)
Level = 500 ug/m^3		1990	1	1	(100)	0	(0)	0	(0)	1	(100)
Level = 750 ug/m^3		1989	2	1	(50)	1	(50)	0	(0)	0	(0)
		1990	3	3	(100)	0	(0)	0	(0)	3	(100)
Level = 1 mg/m^3		1990	2	2	(100)	0	(0)	0	(0)	0	(0)

NOTE: Levels are defined as follows:
Coal Mine Dust Level = 2 mg/m^3 MRE for MSHA coal mine dust sample (level not reduced by quartz content).
Quartz Dust Level = 0.10 mg/m^3 MRE for MSHA coal mine quartz dust sample (2 lpm flowrate).
= 10 mg/m^3 divided by (% quartz + 2) for MSHA metal/non-metal mine quartz dust sample (1.7 lpm flowrate).
= 10 mg/m^3 divided by (% quartz + 2) for OSHA quartz dust sample (1.7 lpm flowrate) (1986-March 1, 1989).
= 0.1 mg/m^3 for OSHA quartz dust sample (March 1, 1989-1990).
Asbestos Fiber Level = 2 fiber/cc (8 hours) and 10 fiber/cc (1 hour) for MSHA metal/non-metal mine asbestos sample.
= 0.2 fiber/cc for OSHA asbestos sample (June 20, 1986-1990).
Cotton Dust Level = 200 ug/m^3, lint free respirable cotton dust in yarn manufacturing and cotton washing operations; 500 ug/m^3, 8 hour TWA, lint-free respirable cotton dust in textile mill waste house operations or lower grade washed cotton in yarn manufacturing; 750 ug/m^3, lint-free respirable cotton dust in slashing and weaving processes; and 1 mg/m^3, in cotton waste processing operations of waste, recycling (sorting, blending, cleaning, and willowing) and garnetting.

Source: Tabulations by Environmental Investigations Branch, DRDS, NIOSH from data tapes provided by OSHA and MSHA.

– indicates data not available.

Table 16. Old Age, Survivors, Disability Insurance (OASDI) Awards for disabled workers with a respiratory diagnosis, by major industry group, 1988 to 1990

Year	Total	Agriculture	Mining	Construction	Manufacturing	Transportation & Public Utilities	Wholesale & Retail Trade	Finance	Services
1988	23,073	7,325	132	407	1,846	614	1,228	263	2,121
1989	21,400	983	572	1,490	5,334	1,988	2,966	586	5,507
1990	22,158	761	455	1,501	5,448	2,033	3,088	622	6,098

NOTE: Because of rounding, components may not add to totals.

SOURCE: Social Security Bulletin, Annual Statistical Supplements.

Table 17. Number of Black Lung beneficiaries and payments by the Social Security Administration and Department of Labor, 1980 to 1990

Year	Social Security Administration		Department of Labor	
	Total beneficiaries	Annual amount (dollars)	Total beneficiaries	Total amount (dollars)
1980	399,477	1,032,000,000	139,073	813,205,000
1981	376,505	1,081,300,000	163,401	805,627,000
1982	354,569	1,076,000,000	173,972	784,035,000
1983	333,358	1,055,800,000	166,043	859,854,000
1984	313,822	1,038,000,000	163,166	873,932,000
1985	294,846	1,025,000,000	160,437	905,517,000
1986	275,783	971,000,000	156,550	829,075,000
1987	258,988	940,000,000	153,289	855,290,000
1988	241,626	904,000,000	149,156	656,689,000
1989	225,764	882,000,000	144,187	650,123,000
1990	210,678	863,400,000	138,491	626,521,000

NOTE: The dollar amounts from the Department of Labor are for fiscal years.

SOURCE: Social Security Bulletin Annual Statistical Supplement 1990 and Black Lung Benefits Act Annual Report on Administration of the Act During Calendar Year 1990.

Table 18. Number of discharges from short-stay non-Federal hospitals, 1970 to 1988

Year	Asbestosis	Coal Workers Pneumoconiosis	Silicosis
1970	300	6,000	6,000
1971	400	8,000	7,000
1972	100	11,000	6,000
1973	2,000	13,000	5,000
1974	1,000	14,000	4,000
1975	1,000	17,000	4,000
1976	1,000	18,000	5,000
1977	1,000	18,000	4,000
1978	3,000	13,000	2,000
1979	3,000	18,000	3,000
1980	4,000	17,000	--
1981	--	14,000	2,000
1982	2,000	17,000	3,000
1983	4,000	22,000	2,000
1984	6,000	23,000	--
1985	6,000	18,000	3,000
1986	6,000	16,000	3,000
1987	11000	17,000	3,000
1988	8,000	15,000	--
1989	8,000	11,000	--

NOTE: No estimates are available for silicosis reports for 1980, 1984, 1988 and 1989. No estimates are available for asbestosis for 1981. Diagnoses are based on ICD-9 codes (see table 1).
NCHS recommends that estimates of less than 5,000 not be used and estimates of 5,000 to 10,000 be used with caution.

SOURCE: National Center for Health Statistics National Hospital Discharge Survey.

ADDENDUM

SENSOR OCCUPATIONAL ASTHMA SURVEILLANCE

Since 1987, the National Institute for Occupational Safety and Health (NIOSH) has funded cooperative agreements with State Health Departments to participate in the Sentinel Event Notification System for Occupational Risks (SENSOR) program. The SENSOR program is a pilot effort to identify occurrences of selected occupational diseases/injuries and to, in turn, provide preventive intervention at worksites targeted as potentially hazardous by this sentinel event surveillance.

Six states (Colorado, Massachusetts, Michigan, New Jersey, New York, and Wisconsin) have been conducting occupational asthma surveillance. The table below shows some occupational asthma data from these states.

Table A1. Number of cases of occupational asthma by state, 1988 to 1991

State	1988	1989	1990	1991	TOTAL
Colorado	31	20	29	22	102
Massachusetts	1	10	10	10	31
Michigan	29	60	117	66	272
New Jersey	22	26	40	41	129
New York	--	15	8	36	59
Wisconsin	--	--	--	--	48

NOTE: Data for Colorado, Massachusetts, New York, and Wisconsin include only cases meeting the surveillance case definition, (see MMWR 1990; 39:121). Data for Michigan and New Jersey include cases of possible occupational asthma (see text below), occupationally aggravated preexisting asthma, and reactive airways dysfunction syndrome (RADS), in addition to cases meeting the surveillance case definition.

SOURCE: JB McCammon, Colorado SENSOR, LK Davis, Massachusetts SENSOR, KD Rosenman, Michigan SENSOR, MJ Stanbury, New Jersey SENSOR, JM Melius, New York SENSOR, and HA Anderson, Wisconsin SENSOR

-- indicates data not available.

OCCUPATIONAL ASTHMA SURVEILLANCE, MICHIGAN

Tables A2 and A3 are from the Michigan SENSOR program. Michigan uses two sources to identify persons with occupational asthma: (1) reports from physicians; and, since 1989, (2) reports from hospitals. In Michigan a person is considered to have occupational asthma from sensitization to a work place exposure if: (1) they have a physician diagnosis of asthma; (2) onset of respiratory symptoms associated with a particular job that then improve or are relieved when the patient is not working; and (3) they work with a known occupational asthmogen, or have evidence of an association between work exposures and a decrease in pulmonary function. If only criteria (1) and (2) above are met the person is considered to have possible occupational asthma. In addition, the Michigan SENSOR program ascertains other categories of asthma associated with work (see footnote, Table A1.) An industrial hygiene investigation at the patient's work site is performed to determine the allergen. If a person had physician diagnosed asthma before beginning work and their asthma became worse at a particular job the person is considered to have aggravated asthma. Occupational asthma from exposure to an allergen at work typically develops after a variable period of symptomless exposure to the sensitizing agent. However, if a person develops asthma for the first time immediately after an acute exposure to an irritating chemical at work the patient is considered to have reactive airways dysfunction syndrome (RADS). After the patient has been interviewed and the work-relatedness of their condition evaluated, an industrial hygiene investigation may be conducted at the patient's work place. At this follow-up investigation, co-workers are interviewed to determine if other individuals are experiencing similar breathing problems from exposure to the suspected asthmogen. An industrial hygienist conducts air monitoring for suspected asthmogen and reviews the company's health and safety program. After the investigation is completed, a report of air sampling results and recommendations for preventing occupational asthma are sent to the company and union.

Table A2. Primary industrial classifications for reported occupational asthma, Michigan, 1988 to 1991

Industry (SIC)	Number of Cases	Percent
Manufacturing		
Automobile (37)	102	37.6
Chemicals and Allied Products (28)	17	6.3
Industrial and Commercial Machinery and Computer Equipment (35)	15	5.5
Rubber and Miscellaneous Plastic Products (30)	15	5.5
Fabricated Metal Products (34)	15	5.5
Foundries (33)	13	4.8
Food and Kindred Products (20)	13	4.8
Miscellaneous Manufacturing (Includes clothing, lumber, paper, and electronics) (22-27, 29, 32, 36, 38-39)	22	8.1
Wholesale and Retail Trade (50, 51, 54, 55, 58)	14	5.2
Services (72, 73, 75, 79, 82, 83, 87)	17	6.3
Health Services (80)	7	2.6
Construction (15, 17)	7	2.6
Miscellaneous (Includes government, utilities, and oil & gas fields) (10, 13, 42, 49, 63, 91-93)	14	5.2
TOTAL	271	100.0

NOTE: For one worker, the industrial classification was not known.

SOURCE: KD Rosenman, Michigan SENSOR

Table A3. Occupational exposures identified for occupational asthma, Michigan, 1988 to 1991

Agent	Number of Cases	Percent
Isocyanates	64	23.5
Unknown (Manufacturing)	48	17.6
Coolant	25	9.2
Unknown (Office)	19	7.0
Formaldehyde	13	4.8
Cobalt	13	4.8
Vehicular exhaust or smoke/fumes	8	2.9
Epoxy	7	2.6
Acrylates	6	2.2
Styrene	6	2.2
Chromium	5	1.8
Flour	5	1.8
Grain Dust	4	1.5
Rose hips	4	1.5
Printing	4	1.5
Other	41	15.1
Total	272	100.0

SOURCE: KD Rosenman, Michigan SENSOR

SILICOSIS SURVEILLANCE

Four states (Michigan, New Jersey, Ohio, and Wisconsin) have been funded to conduct surveillance for silicosis. Each of these states identifies cases of silicosis in a slightly different manner. Michigan seeks out cases from a variety of sources including physician-generated occupational disease reports, death certificates, state Department of Labor files, and hospital discharge data. New Jersey draws cases from these sources as well, excluding Department of Labor files. Ohio includes cases initially identified through death certificates, but only when the diagnosis is confirmed by the patient's physician through follow-up. Wisconsin receives case information only from physician-generated occupational disease reports. States collect demographic, work history and medical information about each silicosis case from case-patient interviews, disease reports, death certificates, hospital records, Department of Labor records, or some combination of these. Silicosis is considered confirmed if: 1) there is a history of occupational exposure to silica; AND 2) a chest radiograph is classified by a NIOSH-certified "B" reader as category 1/0 or greater profusion of small rounded opacities or a lung tissue biopsy report indicates silicosis. Prevention efforts vary among the participating State Health Departments, and include some or all of the following activities: 1) interviews with individuals with reported and/or confirmed silicosis; 2) the distribution of literature regarding the health hazards of silica exposure to confirmed cases and physicians; 3) OSHA or state health department industrial hygiene investigations with environmental monitoring to measure exposures to airborne respirable silica; and 4) referral to appropriate regulatory agencies if excessive exposures or hazardous work practices are found.

Tables A4 through A6 represent a compilation of silicosis surveillance data from four SENSOR states.

Table A4. Number of confirmed cases of silicosis by state and year, through 1990

State	Before 1985	1985	1986	1987	1988	1989	1990	Total
Michigan	9	23	22	83	65	61	40	303
New Jersey	89	15	44	14	45	18	10	235
Ohio	0	0	0	0	0	25	17	42
Wisconsin	0	0	4	17	28	1	4	54
TOTAL	98	38	70	114	138	105	71	634

SOURCE: KD Rosenman, Michigan SENSOR, MJ Stanbury, New Jersey SENSOR, NA Migliozzi, Ohio SENSOR, and HA Anderson, Wisconsin SENSOR.

Table A5. Distribution of confirmed cases of silicosis by duration of exposure to silica by state, through 1990

State	< 10 years	10-20 years	21-30 years	> 30 years	TOTAL
Michigan	26 (8.8%)	53 (18.0%)	96 (32.5%)	120 (40.7%)	295 (100%)
New Jersey	25 (13.7%)	51 (27.9%)	39 (21.3%)	68 (37.2%)	183 (100%)
Ohio	0 (0.0%)	4 (25.0%)	9 (56.3%)	3 (18.8%)	16 (100%)
Wisconsin	3 (6.4%)	16 (34.0%)	24 (51.1%)	4 (8.5%)	47 (100%)
TOTAL	54 (10.0%)	124 (22.9%)	168 (31.1%)	195 (36.0%)	541 (100%)

NOTE: Information not available for 93 cases. Because of rounding, components may not add to totals.

SOURCE: KD Rosenman, Michigan SENSOR, MJ Stanbury, New Jersey SENSOR, NA Migliozzi, Ohio SENSOR, and HA Anderson, Wisconsin SENSOR.

Table A6. Primary industry where silica exposure occurred for confirmed cases of silicosis by state, through 1990

Industry (SIC)	Michigan	New Jersey	Ohio	Wisconsin	Total
Manufacturing					
Primary Metal Industries (33)	241 (80.3%)	44 (19.2%)	8 (28.6%)	37 (77.1%)	330 (54.6%)
Stone, Clay, Glass and Concrete Products (32)	17 (5.7%)	104 (45.4%)	9 (32.1%)	1 (2.1%)	131 (21.7%)
Miscellaneous (22,26,27,28,30,34-38)	21 (7.0%)	29 (12.7%)	10 (35.7%)	8 (16.7%)	68 (11.2%)
Mining (10-14)	10 (3.3%)	30 (13.1%)	1 (3.6%)	2 (4.2%)	43 (7.1%)
Construction (15-17)	6 (2.0%)	16 (7.0%)	0 (0.0%)	0 (0.0%)	22 (3.6%)
Transportation and Communication (42,46-49)	3 (1.0%)	2 (0.9%)	0 (0.0%)	0 (0.0%)	5 (0.8%)
Services (73,76,77,80)	2 (0.7%)	2 (0.9%)	0 (0.0%)	0 (0.0%)	4 (0.7%)
Trade (50,59)	0 (0.0%)	2 (0.9%)	0 (0.0%)	0 (0.0%)	2 (0.3%)

NOTE: Information not available for 29 cases.

SOURCE: KD Rosenman, Michigan SENSOR, MJ Stanbury, New Jersey SENSOR, NA Migliozzi, Ohio SENSOR, and HA Anderson, Wisconsin SENSOR.

APPENDIX

States reporting industry and occupation codes from death certificates to NCHS, 1985 to 1988

1985	1986	1987	1988
—	—	Alaska	Alaska
Colorado	Colorado	Colorado	Colorado
Georgia	Georgia	Georgia	Georgia
—	—	—	Idaho
—	Indiana	Indiana	Indiana
Kansas	Kansas	Kansas	Kansas
Kentucky	Kentucky	Kentucky	Kentucky
Maine	Maine	Maine	Maine
Missouri	Missouri	—	—
Nebraska	—	—	—
Nevada	Nevada	Nevada	Nevada
New Hampshire	New Hampshire	New Hampshire	New Hampshire
—	—	—	New Jersey
—	New Mexico	New Mexico	New Mexico
—	—	North Carolina	North Carolina
Ohio	Ohio	Ohio	Ohio
Oklahoma	Oklahoma	Oklahoma	Oklahoma
Rhode Island	Rhode Island	Rhode Island	Rhode Island
South Carolina	South Carolina	South Carolina	South Carolina
Tennessee	Tennessee	Tennessee	Tennessee
Utah	Utah	Utah	Utah
—	Vermont	Vermont	Vermont
—	—	—	West Virginia
Wisconsin	Wisconsin	Wisconsin	Wisconsin

CROSS-INDEX

Subject	Page numbers
Asbestosis	2, 3, 16, 17, 20, 30
Byssinosis	12, 13, 20
Coal workers' pneumoconiosis	8, 9, 16, 17, 20, 22, 30
Compensation	29
Dust diseases of the lungs	22, 25, 26
Dust exposure levels (coal mine, quartz, asbestos, cotton)	27, 28
Hypersensitivity pneumonitis	14, 15, 20
Occupational asthma	31
Peritoneal malignancies	6, 7, 20
Pleural malignancies	4, 5, 20
Silicosis	10, 11, 16, 17, 20, 30, 33
Toxic respiratory conditions	22